Merzouk Belkacem

Study of soil erosion trends

Merzouk Belkacem

Study of soil erosion trends

using RUSLE, GIS and remote sensing in the Boussellam catchment area in Algeria

ScienciaScripts

Imprint

Any brand names and product names mentioned in this book are subject to trademark, brand or patent protection and are trademarks or registered trademarks of their respective holders. The use of brand names, product names, common names, trade names, product descriptions etc. even without a particular marking in this work is in no way to be construed to mean that such names may be regarded as unrestricted in respect of trademark and brand protection legislation and could thus be used by anyone.

Cover image: www.ingimage.com

This book is a translation from the original published under ISBN 978-620-6-70158-3.

Publisher:
Sciencia Scripts
is a trademark of
Dodo Books Indian Ocean Ltd. and OmniScriptum S.R.L publishing group

120 High Road, East Finchley, London, N2 9ED, United Kingdom
Str. Armeneasca 28/1, office 1, Chisinau MD-2012, Republic of Moldova, Europe
Printed at: see last page
ISBN: 978-620-7-78977-1

Contents

Resume

The aim of this study is to estimate annual soil losses in the Boussellam catchment in north-eastern Algeria, using the Revised Universal Soil Loss Equation (RUSLE), a Geographic Information System (GIS) and remote sensing. The RUSLE model was used to model the main factors involved in the erosion phenomenon.

The Boussellam catchment covers an area of 4,151 km^2 , is elongated and has a low relief. It is characterised by an altitude ranging from 189 m to 1757 m, with an average of 931 m, and a slope varying between 0 and 266%, with an average of 18%. The results show that the erosivity factor R averages 68 (MJ.mm/ha.h.yr) with a maximum value of 140 (MJ.mm/ha.h.yr). The soil erodibility factor K varies from 0.04 to 0.21 with an average of 0.11 (t.h.ha)/(MJ.ha.mm). The topographical factor LS varies from 0 to 216 with an average of 16. The average value of the P factor is 0.77, and that of the C factor is 0.076. The combination of the different maps of these parameters has made it possible to deduce the erosion map, from which it emerges that the erosion phenomenon mainly affects the north of the basin, with an average of 258 (t/ha/yr) for July 2015.

Keywords : *Boussellam catchment, Erosion, ArcGis, RUSLE, Teledetection.*

General introduction

General introduction

Soil erosion is a natural process that is undoubtedly largely responsible for the current gëomorphology. Changes in climate and landscape, under the influence of demographic pressure and the extension of farm crops, have contributed to an increase in the exposure of land to the process of runoff, and consequently to soil degradation through erosion.

Determining the areas at risk of erosion, as well as re-evaluating the factors that control erosion and its characteristics, are complex tasks, but can be resolved with the integration of new methods, such as remote sensing, geographic information systems (GIS) and field surveys.

The aim of this work is to estimate and map erosive risks in the Boussellam basin (Soummam) using revised universal soil loss liquidation (RUSLE), GIS (ArcGis) and remote sensing.

This study is divided into three main chapters:

- The first chapter describes the Boussellam catchment area and its characteristics.
- The second chapter describes the phenomenon of erosion and its process.
- The third chapter is devoted to re-evaluating and mapping the erosion rate.

Finally, the general conclusion summarises the main results obtained and suggests research prospects for the coming years, since the scope of the subject merits much further development.

Description of the study area

Chapter I: Description of the study area

1.1. Introduction

The chapter is devoted to a general description of the catchment ëtudië in order to determine the gëographic, gëomorphological, gëological and hydrogëological characteristics, fundamental tools for understanding hydrological mechanisms. All these characteristics play a key role in the hydrological behaviour of catchments and rivers.

1.2. Geographical location

The Boussellam sub-basin belongs to the large Soummam basin (15) which is located in the central part of northern Algeria (Figs. I.1 - I.3).

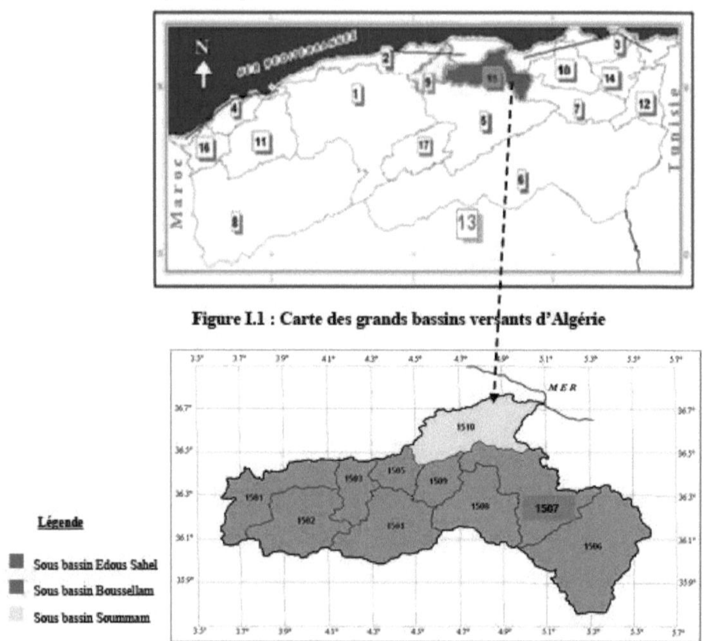

Figure I.1 : Carte des grands bassins versants d'Algérie

Figure I.2 : Les grands sous bassins versants de la Soummam [1]

Figure I.1 : Map of Algeria's major drainage basins
Figure I.2 : The main sub-catchment areas of the Soummam [1].

Figure I.3 : Delimitation of the Boussellam sub-catchment using ArcGis

The Boussellam sub-basin is in turn made up of four (04) sub-basins (1506, 1507, 1508, 1509). This sub-basin has its source in Djebel Meghris to the north of Setif and drains water into the Oued Soummam over a length of approximately 129 km.

The Oued Boussellam is Sëtif's main river. Along with the Oued Sahel to the west, it is one of the two main tributaries of the Soummam, draining 54.9% of the total surface area of the basin to the sea [2]. The surface area of the catchment basin is approximately 4151 km^2 , with a length of 129 km and an annual regularisable volume of almost 38 Hm^3 . It is formed by the combination of the Oued Gassar, which runs along the southern slopes of the Djebel Meghris (altitude 1737 m), and the Oued Ouricia, which lies in the southern part of this Djebel [3].

It rises at an altitude of about 1100 m, five kilometres north-west of the town of Setif. It lies approximately between longitudes 5° 20' 00" and 5° 25' 00" East and 36° 10' 00" and 36° 15' 00" North. The Oued Boussellam passes through several agglomerations in the northern region of Setif (Bougaa, Hammam Gergour, Oued Sebt, Charchar and Beni Ourtillene) and the southern region (Farmatou, Sidi el khier, Mezloug and Hammam Ouled Yelles) (Fig. I.4). Historically, this wadi has always been considered a major wetland by the local population, particularly the inhabitants of the town of Setif. It was used for bathing, fishing, leisure activities and even as a garden [3].

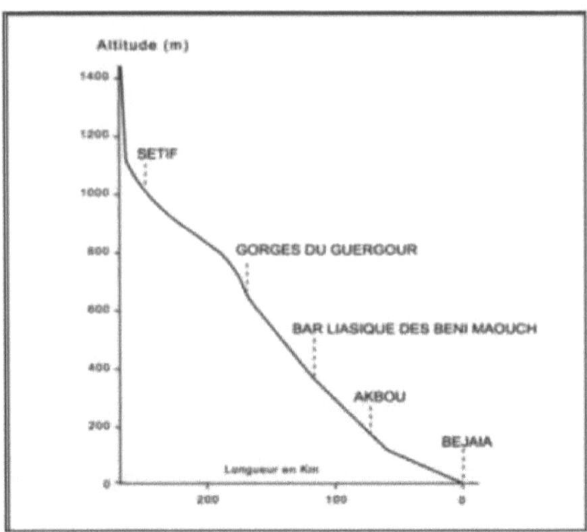

Figure I.4: Longitudinal profile of the Oued Boussellam - Soummam [2].

1.3. Geology

Gëology and lithology are important data for understanding and studying the environment. The nature of the terrain is one of the main criteria determining the choice of development work. From a geological point of view, Algdrie is subdivided into two areas which are opposed in terms of their history and geological structure [4].

- A northern estate in North Algdria.
- Saharan Algeria.

The border between these two domains is delimited by the South-Atlasic accident which follows the southern slope of the Saharan Atlas. Our study area (region) belongs to the first domain and is part of the Maghrebide chain.

1.3.1. Aperqu on the geology of North Algeria

The North African or Maghrebian Alps are part of the Tertiary-age perimediterranean alpine orogeny, which stretches west to east for 2000 km from southern Spain to the Calabrian-Sicilian arc [4].

The North African or Maghrebian Alps are part of the Tertiary-age Pdrimdditerranden Alpine orogeny (*Durand-Delga, 1969*), which extends west to east for 2000 km from southern Spain to the Calabrian-Sicilian arc (Figs. I.5 and I.6).

In Algdria, the Maghrebids range covers the following areas from north to south:

- *the internal domain*: appelë also Kabyle or Kabylide basement, is composedë of mëtamorphic crystallophyllic massifs (gneisses, marbles, amphibolites, micaschists and schists) and a pateozoic sëdimentary ensemble (Ordovician to Carbonifëre) with little mëtamorphism. This basement outcrops from west to east in the Chenoua massif (west of Algiers), the Algiers massif, the Grande Kabylie massif and the Petite Kabylie massif (between Jijel and Skikda). The latter, 120 km long and 30 km wide, is the largest outcrop of the Kabyle basement in Algeria. This bedrock is bordered to the south by the Mësozoic and Cënozoic units of the Kabyle Dorsal, sometimes called the "limestone chain" because of the extent of the limestone Lower Jurassic.

- *the flysch domain*: is made up of sheets of crëtacës-palëogënes flysch which outcrop in coastal areas 800 km long, between Mostaganem and Bizerte (Tunisia). They are essentially deep-sea detritus deposited by turbid currents. These flyschs are presented in three ways:
- *i*) in an internal position, superimposed on the Kabyle massifs, i.e. retro-charriged on the internal zones, and known as the North Kabyle flyschs;
- *ii*) relatively outside the southern edge of the Kabyle Dorsal (South Kabyle flysch)
- *iii*) in a very external position, in the form of isolated masses floating on the Tell charri'es up to a hundred kilometres to the south.

Figure I.5 : The perimediterranean alpine orogeny [5].

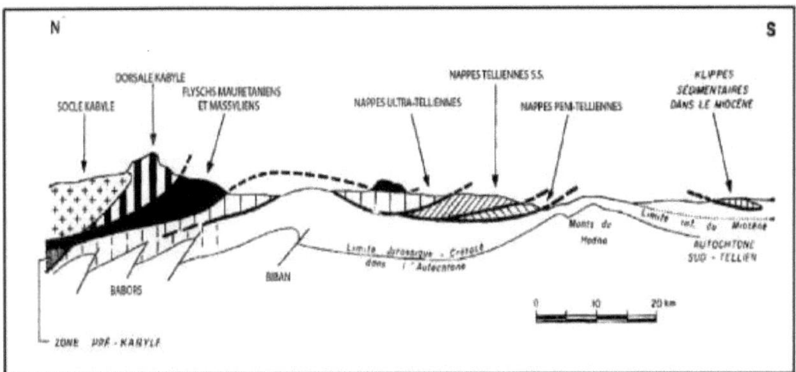

Figure I.6: Structural relationships between the different units of the Maghreb charne [5].

II.3.2. Lithology of the formations

The geology of the Oued Boussellam basin has been ël.ëtudiëe by rëfëranting old geological works: the gëologique map of Sëtif at 1/50 000, the gëologique map of A^rie at 1/500.000.00. Oued Boussellam stretches over limestone rocks belonging to the Quaternary [4]. From a lithological point of view, we can distinguish limestone rocks belonging to the Quaternary, recent alluvial deposits consisting of Quaternary dëpôts of lacustrine limestone origin, and Villafranchian crusts with sphëroideous stony horizons. The Mio-pliocene shows the presence of fluvio-lacustrine dëp6ts usually offering a reddish colouration, quite pronouncedëe, consisting of sands, gravels silt and clays.

The bed of the wadi is bordered by the Tellian formations, which are mainly represented by upper matërial ëocëne units and are located on the south-western slopes of Djebel Megress in addition to the middle and upper Ocene &тë by black, brown and grey marls. The southern part of the Sëtif geological map is made up of the Djemila nappe, which is composed of white bituminous limestone with a black fracture and black flint. To the north, there are limestone formations consisting of black marly rocks and grey, white or black massive limestone (Fig. I.7) [1].

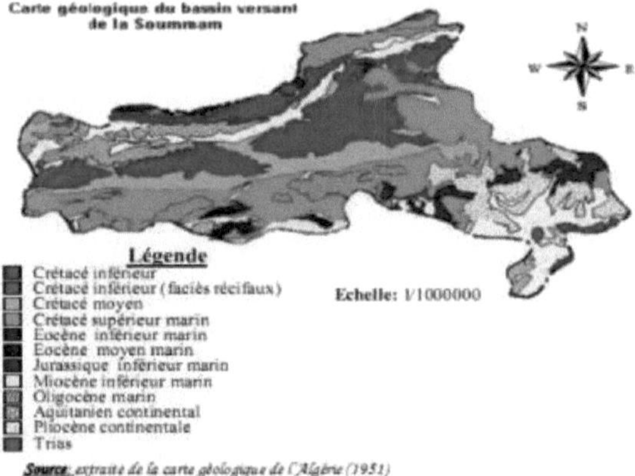

Figure I.7 : Geological map of the Soummam catchment area [1].

1.4. Soils

In the Soummam watershed we find ancient soil types of palëo- marecageous formation which are caractërisës by well accentuated formations. The majority of soils in the Soummam basin are calcareous soils (rich in limestone). These soils generally have a tegere texture and are therefore permëable (Fig. I.8).

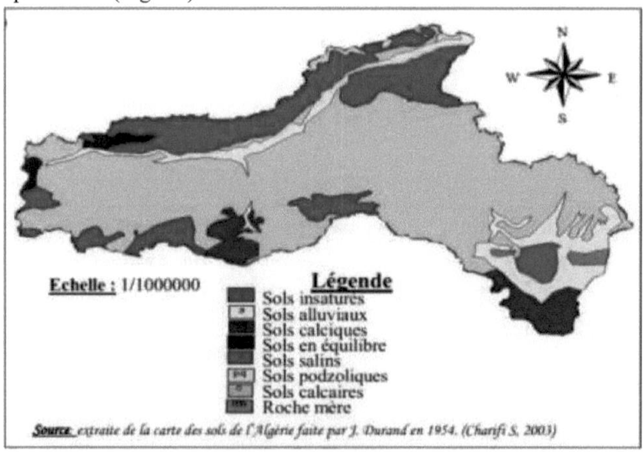

Figure I.8: Soil map of the Soummam catchment area [6].

In the north of the basin, we find unsaturated soils that do not contain limestone, the clay of which may be more abundant at the surface than at depth; their bedrocks are generally impermëable or give products of impermëable dëcomposition. Along the Oueds, the existing soils are alluvial dëp6ts called alluvial soils. In addition, there are small quantities of calcareous soils and soils in equilibrium that are not very thick, more or less rich in limestone and very poor in soluble salts.

1.5. Plant cover

Vegetation is an important physical parameter in catchment areas. Depending on its nature, diversity and density, vegetation cover has a direct influence on the water cycle within catchments. Vegetation evolves according to the climatological conditions of the environment. Vegetation density decreases with altitude as a result of changes in climatological conditions. If vegetation density decreases, then evapotranspiration and interception losses decrease from downstream to upstream.

On the Akfadou, Timezrit and Babors massifs, the vegetation is dense and varied, forming vast forests of oak, cork oak and Aleppo pine. On the other hand, it is less dense and localized on the intermediate reliefs, or even absent on the marly hills. It is represented by dwarf oak and lentisque olive scrub. In addition, the mountainous zone remains a region of arboriculture, notably olive and fig trees. On the other hand, in the depressions and given the saline quality of the soil, the flora is generally poor [2]. On the plain, the vegetation is dense but essentially temporary; it is formed by the large and formidable fields of various vegetable crops. In the past, the forests that covered the region provided the wood needed for a flourishing timber industry, but unfortunately this capital is tending to disappear under the fires that ravage thousands of hectares every year. Added to this is the absence of a clear reforestation and fire-fighting policy [1].

1.6. The river system

The Soummam catchment area is highly developed hydrographically by its various watercourses. Most of its waters are drained by the Oued de Boussellam and Oued Sahel, which meet near Akbou to form the Oued Soummam (Fig. I.9).

In most of the Soummam catchment area, groundwater is confined in deep layers and can only be brought to the surface by upward currents [6].

The hydrographic network of the Boussellam sub-basin plotted by ArcGis is shown in Figure (I .10).

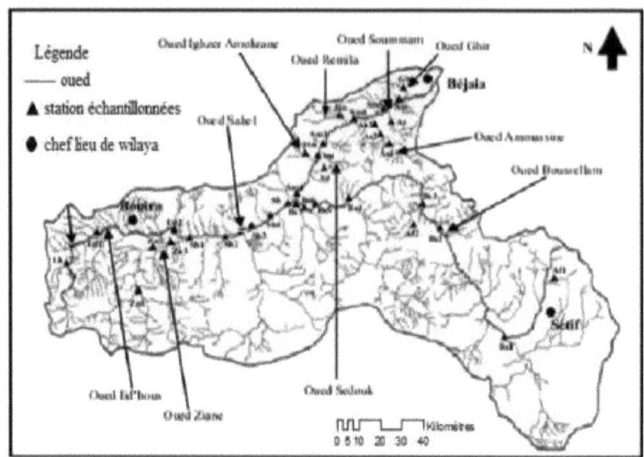

Figure I.9: Hydrographic network of the Soummam basin [6].

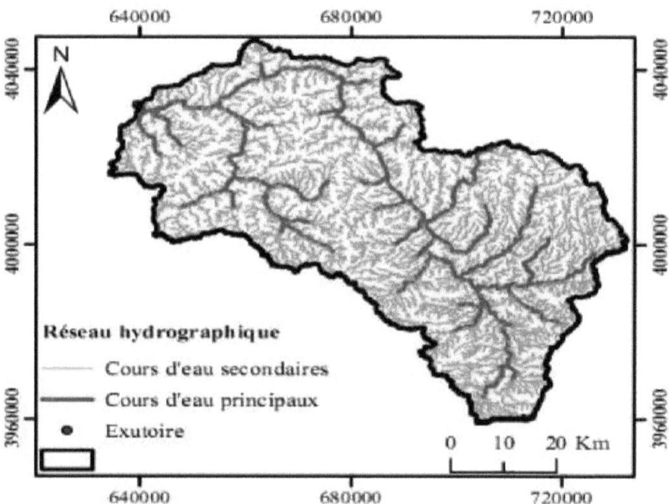

Figure I.10: Hydrographic network of the Boussellam sub-basin

1.7. Climate situation

1.7.1. Climate

The Soummam region is subject to three types of climate: a temperate coastal climate on the lower reaches of the Oued Soummam, a Tellian Atlas climate on the middle Soummam and part of the upper Soummam basin (Oueds Sahel - lower reaches of the Oued Boussellam) and a high plains climate on the upper reaches of the Oued Boussellam basin. A bioclimatic ëtages map is representedësentëe in figure (I.11). According to the latter, the climate of Boussellam is humid to sub-humid in the eastern and northern regions, a semi-arid climate in the centre and an arid climate in the south.

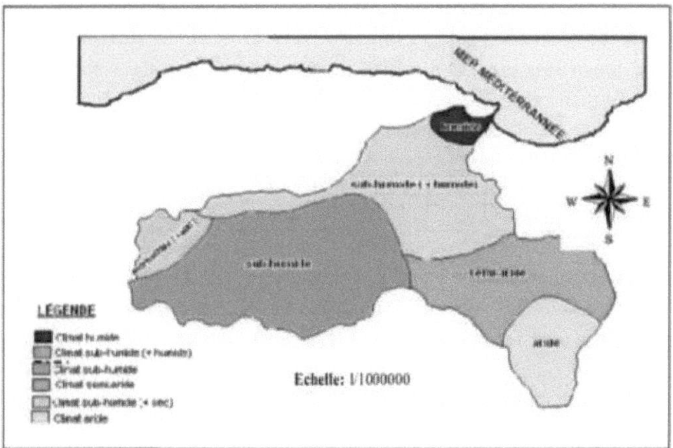

Figure I.11: Bioclimatic stages in the Soummam basin [1].

1.7.2. Temperature

As an indication, the drop in maximum temperatures is 0.65°C for a ëlë elevation of 100 m, and the drop in minimum temperatures is of the order of 0.4°C for the same ëlë elevation. The maximum altitude in our study area is around 1757 m and the minimum altitude is 189 m. This means that the average altitude of our study area is around 973 m. The average monthly tempëratures recorded from 1978 to 2017 are shown in the table below.

Table I.1: Adjusted mean monthly temperatures in degrees Celsius for the study region

Month	Jan.	Feb.	March	Apr.	May	June	Jul.	August	Sep.	Oct.	Nov.	Dec.
Tmax (C°)	11.75	11.62	13.28	15.01	17.53	21.03	24.13	24.88	22.79	20.07	15.42	12.27
T min (C°)	4.14	4.33	5.69	7.49	10.63	14.33	17.33	17.91	15.8	15.8	8.44	5.33
Tmoy (C°)	7.64	7.97	9.48	11.25	14.08	17.68	20.73	21.39	19.29	16.31	11.93	8.8

With :

Tmax: Maximum temperature, Tmin: Minimum temperature

Tmoy: Average temperature. $[T°_{moy} = (T_{max} + T_{min})/2]$

January was the coldest month, with a temperature of 4.14°C, while July and August were the hottest, with values of 24.13°C and 24.88°C.

1.7.3. Wind

It is a factor that favours evaporation, transporting saturated layers of air that are close to the surface of the water or soil to be replaced by layers of air that are more or less dry.

In the Soummam region, the prevailing winds are from the north-west (sea breezes which easily blow into the Soummam valley). In winter, they bring clouds, rain. In winter, the south-east winds are much less important, especially the south wind, the sirocco, a dry wind of variable intensity that has a disastrous effect on vegetation. The monthly averages of wind speeds over ten years are given in the table below.

Table I.2: Average monthly wind speed (1994-2003) [1].

Month	Jan.	Feb.	March	Apr.	May	June	Jul.	August	Sep.	Oct.	Nov.	Dec.
v (m/s)	4,7	4,28	3,92	3,52	3,58	3,47	3,63	3,38	3,54	4,13	4,4	4,91

The lowest wind speed was recorded in August, at 3.38 m/s, in contrast to the winter period when the winds are cold and violent, especially in December and January, with wind speeds varying between 4.91 and 4.7 m/s.

1.7.4. Precipitation

Two groups of factors - geographical (*distance from the sea, altitude, exposure of slopes to north-westerly rainy winds*) and meteorological (*movement of cold, wet oceanic polar air masses, warm, wet tropical air masses from the South Atlantic and finally continental tropical air masses or the Saharan anticyclone*) - influence the spatial distribution of precipitation, as well as the structure of rainfall patterns [1].

The temporal irregularity of rainfall is a fundamental feature of the Algerian climate. The rainfall map of North Algeria drawn up by the Agence Nationale des Ressource Hydrique (ANRH - Project PNOD/ALG/88/021-January 1993) provides the mean annual rainfall for the period 1969 - 1989. The following points can be observed:

- Maximum rainfall on the Djurdjura massif of 1000 mm to 1500 mm.
- The Sahel basin, in the centre and south, has rainfall of between 300 and 500 mm.
- The downstream basin, near Bejaia, is subject to rainfall of 600 to 900 mm.
- The Boussellam basin comprises two sectors: the south, characterised by an arid climate, has rainfall of less than 400 mm, or even 300 mm. The centre and north of the basin have rainfall of more than 500 mm, with average values of around 600 mm, and up to a maximum of 700 mm.

The p1иуютёlпе of the Soummam watershed is well dëterminëe thanks to the existence of around 50 p1uviomëtric stations, i.e. approximately one station for every 180 km^2 . Rainfall data from thirteen (13) stations located in the Boussellam basin show that the average annual rainfall is around 423.31 mm [Table I.3].

Table I.3 : Monthly and annual precipitation values for 13 stations in the Boussellam sub-catchment (1970 -2017) [4,6].

N°	Code	Seven	Oct	Nov	Dec	Jan	Feb	Mar	Apr	May	June	July	August	P' an
1	150603	27.78	32.42	34.35	26.47	31.27	30.07	31.78	34.73	38.66	20.56	7.9	10.13	346.4
2	150606	32.95	36.72	37.18	63.73	48.28	39.9	43.92	40.81	43.43	21.81	8.35	11.11	420.32
3	150607	33.29	37.74	39.38	54.83	49.55	41.49	45.86	43.36	44.3	22.06	8.39	11.22	432.45
4	150608	32.81	36.88	39.5	60.29	52.54	46.1	51.62	46.77	43.59	20.85	7.87	10.75	441.99
5	150614	28.3	34.36	37.18	40.26	38.19	33.52	33.89	37.62	40.01	21.32	8.02	10.43	370.9
6	150703	27.52	47.98	57.82	98.92	92.93	84.26	65.88	57.64	43.85	15.81	4.81	7.84	612.82
7	150706	36.07	46.51	52.04	81.51	70.6	61.89	62.65	52.45	49.13	20.93	7.89	11.22	555.94
8	150707	29.51	31.73	39.07	50.84	39.87	36.72	42.53	37.81	39.8	20.59	8.32	9.06	385.85

9	150802	31.79	39.72	48.14	80.6	61.97	57.97	59.23	55.47	45.3	19.79	7.16	10.37	500.62
10	150904	26.31	32.83	43.04	62.16	56.59	42.31	42.82	38.37	29.24	8.14	4.49	8.55	394.85
11	150602	29.51	31.73	39	50.84	39.87	36.72	42.53	37.81	39.8	20.59	8.32	8.5	385.22
12	50901	26.6	28.8	35.27	42.85	47.47	32.47	28.67	33.3	29.8	7	5.4	4.19	321.82
13	150801	33.12	26.89	31.3	38.47	36.56	27.24	31.12	39.59	38.54	15.94	3.97	11.44	333.88
											Annual average			**423.31**

1.8. Transport of suspended solids

The estuary of the Soummam catchment area receives a large proportion of the volume of urban and industrial wastewater, in addition to the wastewater from upstream.

The qualitë of the water in the estuary is clearly dëgradëe, given that the concentrations of suspended solids (SS) are very high and far exceed the standards for surface water (> 50 mg/L), in addition to a load of dissolved solids expressed by fairly high electrical conductivities of between 1.26 and 13.22 mS/cm. Dissolved oxygen (DO) concentrations are relatively low (2.82 - 5.4 mg/L). A clear temporal variation in these parameters (TSS, Conductivity, DO) was observed during the përiod of study, this being due mainly to variations in precipitation rates (natural factor), in addition to urban and industrial discharges whose volume is very irregular over time (anthropogenic factor) [7].

1.9. Morphometric characteristics

The morphological parameters of a catchment (shape, altitude, slope, relief, etc.) play an essential role in its hydrological behaviour. They have the advantage of being amenable to quantified analysis, which should be as precise as possible from the outset of any study. The shape of the catchment area, which can be expressed in terms of the *Gravelius* compactness index, also has a definite influence on the flow.

The results of this ëtude have ël.ë rëalisës using ArcGis 10.1 software.

1.9.1. Surface area (A)

As the catchment area is where rainfall falls and watercourses are fed, the dëbits are partly based on its surface. The surface area of the catchment area can be measured by superimposing a grid drawn on transparent paper, by using a planimeter or by digitisation techniques (Arc Gis, Global Mapper, MapInfo software, etc.). The surface area of the Boussellam catchment is 4150.9 km^2.

1.9.2. The perimeter

The përimëtre represents all the irregularities in the contour or boundary of the catchment area. It is expressed in km. The watershed contour is formed by a line joining all the highest points. It has no influence on the state of flow of the watercourse at catchment level. The perimeter can be measured using a scaler or automatically using the software mentioned above. The përimëtre of the B.V of Boussellam is 421.41km.

1.9.3. The shape

The most widely used index for determining the shape of a catchment is the *Gravelius* Compaction Index (KG). The shape of a catchment has a major influence on the overall flow of the watercourse, and especially on the shape of the hydrograph at the catchment outlet, resulting from a given rainfall. It is established by comparing the përimëtre of the basin (P) with that of a circle with the same surface area. This index is given by the following relationship (Roche, 1963) [8] :

$$K_G = \frac{P}{2\sqrt{\pi A}} = 0.28 * \frac{P}{\sqrt{A}}$$

- If K_G is close to **1**, the catchment area is almost circular - If K_G is greater than **1**, the catchment area is elongated.

With :

K_G: *Gravelius* compaci index

P: Përimëtre of catchment area in km

A: Catchment area in km^2 .

$$K_G = 1.83 \Rightarrow$$ the Boussellam catchment is аПопдёе in shape.

1.9.4. The equivalent rectangle

This is the rectangle of length **L** and width l which has the same surface area and the same irerimeter as the catchment area. The rectangular watershed is the result of a gëomëtric transformation of the real watershed in which the same surface area, the same përimetre (or the same coefficient of compactness), and therefore consequently the same hypsomëtric distribution are retained. The contour lines become straight lines parallel to the small cötë of the rectangle. Climatology, soil distribution, vegetation cover and drainage density remain unchanged between the contour lines.

In 1963, *Roche* ë-established two relations relating to the dimensions of the ëquivalent rectangle, whose dimensions are given by the following relations:

$$L_r = \frac{K_G \sqrt{A}}{1.12} \left[1 + \sqrt{1 - \left(\frac{1.12}{K_G}\right)} \right]$$

$$\ell_r = \frac{K_G \sqrt{A}}{1.12} \left[1 - \sqrt{1 - \left(\frac{1.12}{K_G}\right)} \right]$$

With :

Lr : Length of the ëquivalent rectangle in km

ℓ_r : Width of rectangle ëquivalent in km

K_G: Compact index

A: Catchment area in km^2 .

In our case, we obtain :

L_r = 188.72 km   **ℓ_r = 21.99 km**

1.9.5. Altimetry and hypsometric curve

The topography of the basin plays a very important role in hydrology, conditioning runoff, infiltration and evaporation.

The results of this application are shown in the following tables.

The altimëtric frequencies are represented in Table (I.4) and by the hypsomëtric curve (Fig. I.12).

These representations show that the higher altitude class (1364 - 1757 m) represents only 55.75 Km^2 , or approximately 1.34 % of the total surface area of the basin. 24.81 % of the surface area of the basin lies between 860 and 945 m altitude, with a surface area of 1029.752 km^2 .

Table I.4: Breakdown of different altitude bands

Altitude range	Surface area (km)²	Surface (%)	Surface Cumulative (%)	Surface Cumulative (km2)	Altitude (m)
189 - 441	109.0739	2.63	2.63	109.0739	189
442 - 600	200.0107	4.82	7.45	309.0846	442

601 - 742	285.8429	6.89	14.34	594.9275	601
743 - 859	433.6908	10.45	24.79	1028.6183	743
860 - 945	1029.752	24.81	49.6	2058.3703	860
946 - 1019	923.6588	22.25	71.85	2982.0291	946
1020 - 1111	564.8208	13.61	85.46	3546.8499	1020
1112 - 1222	374.721	9.03	94.49	3921.5709	1112
1223 - 1363	173.2392	4.17	98.66	4094.8101	1223
1364 - 1757	55.75092	1.34	100	4150.56102	1364
Total	4150.56102	100			1757

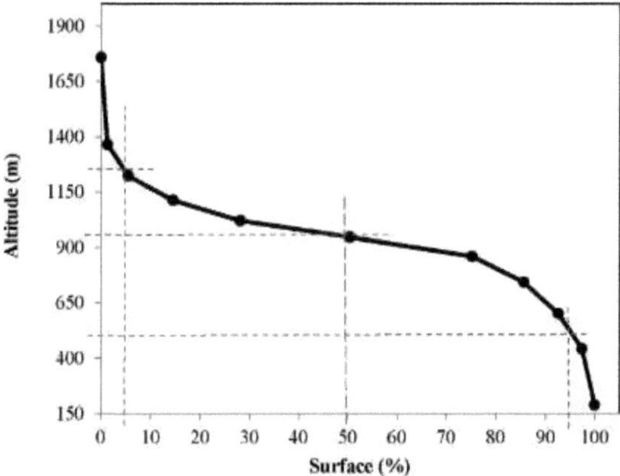

Figure I.12 : Hypsometric curve of the Boussellam catchment area

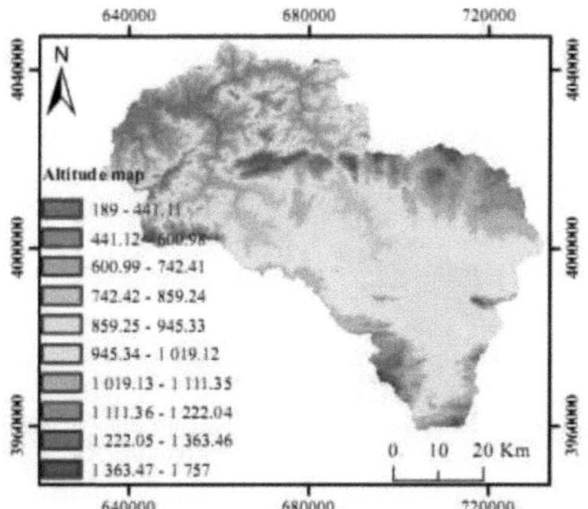

Figure I.13: Map of altitude ranges in the Boussellam basin

I.9.6. Characteristics of altitudes

1) Average altitude (*Hmoy*)

Mean altitude is not very representative of the rёaШё. However, it is sometimes used in the revaluation of certain hydromёtёorological parameters or in the implementation of hydrological modёles. It can be derived directly from the hypsomёtric curve or from reading a topographic map. It can be defined as follows:

$$H_{moy} = \frac{\sum A_i \, H_i}{A_t}$$

With :

H_i : Average altitude between two contour lines (m)

A_i: Surface area ëlëmentaire between two contour lines (Km)^2

A_t: Total surface area of the basin (Km).^{2}

In our case, Hmoy = 931.65 m.

2) Median altitude (H_{50})

Corresponds to the 50% point on the hypsometric curve, H50% = 940 m.

3) Extreme altitudes

The main extreme altitudes are as follows:

- The minimum altitude is: Hmin= 189 m
- The maximum altitude is: Hmax=1757 m
- The height at 5% of the total surface area is: H_5 % = 1250 m
- The height at 95% of the total surface area is: H_{95} % = 520 m.

4) Denivelee simple (D)

On the hypsomëtric curve tracëe, we take the vertical distance in (m), which sëpare the altitudes having 5% and 95% of the total surface of the subcatchment.

$$D = H5\% - H95\% = 730 \text{ m.}$$

5) 10. Slope indices

The aim of these indices is to characterise the slopes of a catchment area and enable comparisons and classifications to be made. They include :

5.1.1. Overall slope index (I_g)

The overall slope index is determined from the hypsometric curve by eliminating the extreme values, so as to retain only 90% of the sub-basin area. This index is calculated using the following formula:

$$I_g = D / L_r$$

With :

D: Denivelee simple (m)

Lr : Length of equivalent rectangle (km) ;

$$\Rightarrow I_g = 3.86 \text{ m/km} = 0.00386$$

$I_g < 0.002$		relief	very low		
$0.002 <_{Ig}$	<0	.005	relief	low	
$0.005 <_{Ig}$	<0	.010	relief	fairly low	
$0.010 <I_g$	<0	.020	relief	moderate	
$0.020 <I_g$	<0	.050	relief	fairly strong	

According to the ORSTOM (*Office de Recherche Scientifique de Territoire d'Outre-Mer*) relief classification, I_g is between 2 and 5 m/km, so the Boussellam sub-basin has little relief.

5.1.2. Specific Level (SL)

The specific denivelee gives an indication of the relief according to the **ORSTOM** classification (Table I.5). The I_g index decreases for the same basin as the surface area increases, so it is difficult to compare basins of different sizes, hence the need to introduce the parameter (DS).

$$\mathbf{D_s = I_g * [A]^{1/2} \implies D_s = 249.21\ m}$$

According to the second ORSTOM classification, the Boussellam basin's spëcific drainage is in class R5, characterised by a fairly high relief.

Table I.5: ORSTOM classification [8].

R1	Very low relief	$05 < Ds < 10$ m
R2	Low relief	$10 < Ds < 25$ m
R3	Fairly low relief	$25 < D_s < 50$ m
R4	Moderate relief	$50 < D_s < 100$ m
R5	**Fairly strong relief**	**$100 < Ds < 250$ m**
R6	Strong relief	$250 < Ds < 500$ m
R7	Very strong relief	$500 < D_s < 750$ m

The Boussellam catchment has the following slope map.

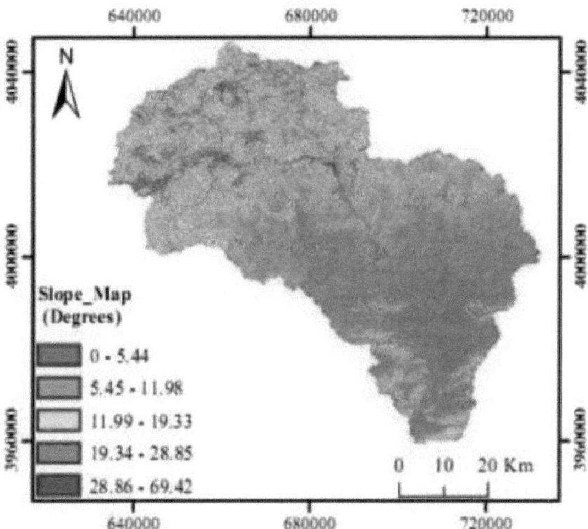

Figure I.13: Slope map of the Boussellam basin

The map shows that over 84% of the surface has a slope of less than 19°, 11% represents 28.85° and around 4% has a slope of 69.42°.

1.11. Characteristics of the river system

1.11.1. Drainage density (Dd)

Drainage density depends on the gëology (structure and lithology), the topographical characteristics of the catchment area and, to a certain extent, climatic and anthropogenic conditions. In practice, drainage density values vary from 3 to 4 for regions where the flow has only reached a very limited development, and is centralised; these values exceed 1000 for certain areas where the flow is very branched with little infiltration.

According to *Schumm,* the inverse value of the drainage density (**C = 1/Dd**) is called the "stream stability constant". Physically, it represents the surface area of the basin required to maintain stable hydrological conditions in a unit hydrographic vector (section of the network). Drainage density® is defined by [6,8]:

$$D_d = \frac{\Sigma L_i}{A} \quad \text{......... (km/km}^2\text{)}$$

Or,

ΣL_i: cumulative length of all the thalwegs in the basin in km.

A: basin area in km^2 .

$$\Rightarrow \quad D_d = \frac{51260.787}{4150.9} = 12.34 \text{ km/km}^2$$

1.11.2. Hydrographic density (F)

Hydrographic density represents the number of flow channels per unit area. It is given by the relationship :

$$F = \frac{\Sigma N_i}{A} \quad \text{......... (km}^{-2}\text{)}$$

N_i: number of watercourses.

$$\Rightarrow F = 10081/4150.9 = 2.42 \text{ km}^{-2}.$$

1.11.3. Coefficient of torrentiality (CT)

$$C_T = \frac{N_1}{A} * D_d$$

$$\Rightarrow \quad C_T = 14.8$$

N_I: number of order 1 rivers (N1 = 4975).

1.11.4. Concentration time (CT)

The time of concentration is defined as the time it takes for a drop of water falling in the furthest point of the catchment area to reach the outlet. It is expressed by *Giandotti*'s formula [8].

$$T_c = \frac{4\sqrt{A} + 1.5 * Lp}{0.8\sqrt{H_{moy} - H_{min}}} \quad \text{...... (h)}$$

Or,

A: surface area of the basin (km2)

Lp: length of main embankment (km)

H_{moy}: average altitude (m)

H_{min}: minimum altitude (m).

$$\Rightarrow T_c = 20,70 \text{ h.}$$

1.11.5. Water flow velocity (ve)

It is given by the following expression.

$$v_e = \frac{L_p}{T_c} \quad \dots \text{(m/s)}$$

$$\Rightarrow \quad V_e = \frac{129}{74520} = 1.73 \text{ m/s}$$

The following table summarises the main characteristics of the ëtudië basin.

Table I.6: Main morphometric characteristics of the Boussellam basin

Parameters	Symbol	Unit	Value
Area	A	Km²	4150.9
Perimeter	P	km	421.41
Compaction index	KG	/	1.83
Maximum altitude	Hmax	m	1757
Minimum altitude	Hmin	m	189
Average altitude	Hmoy	m	931
Overall slope index	Ig	%	0.0038
Width of equivalent rectangle	lr	km	21.99
Length of equivalent rectangle	Lr	km	188.72
Denivelee	D	m	730
Specific level	Ds	m	249.21
Altitude at 5%.	H5%	m	1250
Altitude at 95%.	H95%	m	520
Total drainage density	Dd	km/km²	12.34
Concentration time	Tc	hours	20.70
Horton classification	order	/	7
Hydrographic density	F	1/km²	2.42

1.12. Partial conclusion

In this chapter we have triedë to represent the Boussellam catchment area, to know the topography, the climatic conditions and the current state of the region's hydrographic network.

This basin has a surface area of approximately 4,151km^2 , a 421km diameter, an average altitude of 931m, low relief and an elongated shape. The Oued Boussellam, along with the Oued Sahel in the west, is one of the two main tributaries of the Oued Soummam.

The Boussellam basin is characterised by a humid to sub-humid climate in the east and north, a semi-arid climate in the centre and an arid climate in the south. Precipitation is less than 300 mm in the south and more than 500 mm in the centre and north.

Knowledge of the study region is very important before any in-depth study of the basin (hydrology, erosion, solid transport, flooding, etc.). In this study we will focus on the phënomëne of erosion.

The phenomenon of erosion

Chapter II. The phenomenon of erosion

11.1. Introduction

Soil erosion is a natural or gëological phëne in which soil particles are dëtachëed and dëplacëed by several main factors which are: rainfall, vëgëtation, soil, gëomorphology (slopes in particular) and the impacts of human land use. Erosion can take different forms that combine in time and space: sheet or gully erosion and linear or gully erosion, and it has serious consequences for our environment and our activities and can cause major damage.

Measures of soil loss are expressed locally as the net mass of soil lost over a given period for a given surface area (plot scale). Sediment production is defined as the mass of sediment released by a catchment of any size over a given period.

11.2. Definitions

- Soil erosion, the movement of soil from one place to another, occurs as a result of three main phenomena. It occurs naturally on agricultural land through the action of wind and water, and can be accelerated by certain agricultural activities (fallow land, row crops). It is also caused directly by the tillage method, which results in the progressive movement of soil down slopes, leading to soil loss at the top and accumulation at the base of slopes [9,10].

- Soil erosion is a common term that is often confused with soil degradation as a whole, but in fact refers only to the absolute loss of soil in terms of topsoil and nutrients. This is the most visible effect of soil degradation, but it does not cover all its aspects. Soil erosion is a natural process in mountainous areas, but it is often amplified by poor management practices [10].

11.3. Factors influencing erosion

In addition to gëological and topographical factors, a number of other agents have a direct or indirect influence on the erosion process in catchment areas. Torrential rain is the main agent of the phenomenon, and rainfall irregularity is linked to very high intensities that can cause considerable soil losses. The multiplicity of invoices that influence erosion means that we need to be aware of their direct effects on the erosion process.

11.3.1. Influence of climate

Climate is an important factor influencing water erosion through the nature of rainfall and temperatures. An aggressive climate such as that of arid and semi-arid regions is characterised by heavy rainfall over a short period of time, and long periods of drought. The two elements accentuate erosion to a remarkable degree: high temperatures dry out and crack the soil, making it more vulnerable to erosion, and heavy rainfall uses high levels of kinetic energy to loosen and carry away a maximum amount of soil that is already very weak.

11.3.2. Influence of the soil

It is characterised by its type, texture and condition. In the dry season, soil moisture is almost non-existent, which encourages water erosion during the first rains. It also influences runoff through its infiltration and retention capacities.

11.3.3. Topographical factors

The essential topographical factors are the slope of the catchment, the relief, the density of drainage, the importance of valleys and floodplains, the orientation and size of the catchment. Steep slopes with rapid run-off are generally the cause of excessive erosion, the extent of which depends on the geology of the soil and the protection of the plant cover.

11.3.4. Geological factors

These are factors relating to surface rocks. If the rocks are exposed to rain, wind and gravitational forces, the granulometric distribution of the soils may deteriorate, their permeability, and the presence of certain chemical elements and organic matter affect the erodibility of the soils.

11.3.5. Vegetal convert

Good vegetation cover limits erosion by dissipating rain energy. It favours infiltration and opposes all forms of erosion.

11.3.6. Unsuitable ploughing

Tillage has two opposing effects on water erosion. On the one hand, it limits erosion by increasing the permeability and retention capacity of the soil, and on the other hand, it promotes erosion by reducing the cohesion and structural stability of the soil during low rainfall (or the first phase of heavy rainfall). Proper ploughing considerably reduces water erosion.

11.4. Types of erosion

11.4.1. Wind erosion

Wind exerts pressure on solid particles at rest on the surface exposed to the air flow, applied above the centre of gravity, which is opposed by friction on the base of the particles. These two forces form a torque that tends to cause heavy particles (0.5 to 2 mm) to tip and roll, and the difference in speed between the base and the top of the particles causes them to be sucked upwards. The lighter particles rise vertically until the velocity gradient no longer carries them. They then fall back down, pushed by the wind, following a sub-horizontal trajectory. As they fall, these grains of sand transmit their energy to other grains of sand (as in a game of balls) where the silty-clay aggregates degrade, releasing dust [11].

Wind erosion is most prevalent in West Africa in the dry tropics, where annual rainfall is less than 600 mm, the dry season lasts more than six months and steppe-like vegetation leaves large patches of denuded soil. Elsewhere, it can also develop in soil preparation conditions that lead to heavy spraying of dry surface materials.

11.4.2. Artificial or dry mechanical erosion

This phenomenon of erosion is not caused by water; it is the tillage of the soil that tears up the particles, transports them and deposits them either at the bottom of the plot or on the embankment.

11.4.2.1.   Mechanical erosion factors

The factors influencing the amount of soil moved are :
- The type of tool
- Frequency of visits
- Slope. The steeper the slope, the more the clods of soil roll downwards. This explains why hilltops are often décapés.

11.4.2.2.   Tillage orientation

This can be oriented either along contour lines, or from the top of the plot downwards (which is the case for tractors working on slopes greater than 15%), or from the bottom of the plot upwards (for manual work in particular). It is very rare for the soil to be lifted up by the tools. On the other hand, in the mountains and in areas where soil is scarce, soil is sometimes collected mécanically or in small baskets from the plains and brought up into the mountains, as is the case with the vines. It has also been found that the to-and-fro movement of tools can considerably reduce the speed of removal by mécanic erosion [12].

11.4.3. Water erosion

Water erosion is a complex phёnomёne, which ipirticularly threatens water and soil potentials. It is defined as the detachment and transport of soil particles from its original location by various agents towards a place of dëp6t. Therefore, the three stages through which erosion takes place are detachment, transport and sedimentation. However, it should be pointed out that rainfall and surface runoff are responsible for the detachment, transport and deposition of the soil particles removed. Hydric soil erosion can be defined as the detachment and translocation of soil particles by water deplagating them from their original location to new areas of deposition. Soil erosion is commonly recognised by the incisions or sedimentations that form on the earth's surface. Soil water erosion can be defined as the phenomenon whereby the soil loses some or all of its particles under the action of water [10].

11.4.3.1.   Mechanism of water erosion

The main mechanisms leading to water erosion are :

a)   Secondment

a.1) Moistening due to the impact of raindrops

The four processes that can be identified as responsible for disaggregation are :

- *Bursting,* corresponding to desaggregation by compression of the trapped air during wetting (Fig. II.1). The intensity of bursting depends, among other things, on the volume of trapped air, and therefore on the initial water content of the aggregates and their porosity.

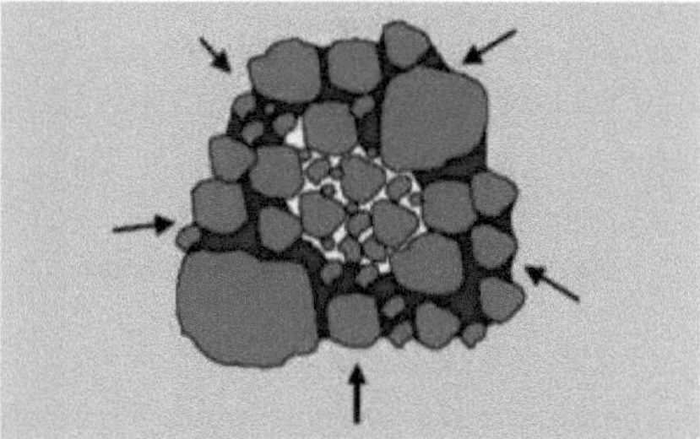

Figure II.1: Break-up of aggregates

- *Differential swelling.* This phĕnomĕne occurs as a result of the moistening and drying of the clays, causing cracks in the aggregates. The extent of this mĕcauisme dĕpend largely on the content and nature of the clay in the soils.
- *Physico-chemical dispersion.* This corresponds to the reduction in the forces of attraction between colloidal particles during wetting. It depends on the size and valency of the cations (particularly sodium) that can bind the negative charges in the soil.
- *Mechanical disaggregation* under raindrop impact (= splash disaggregation) (Fig. II.2). The impact of raindrops can fragment aggregates and, above all, detach particles from their surface.

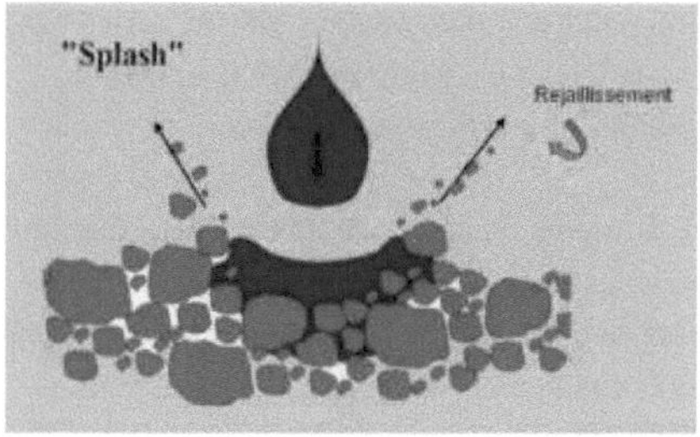

Figure II.2 : Splash detachment

a.2) **Runoff**

Soil erosion occurs when rainwater, no longer able to infiltrate the soil, runs off over the plot, carrying away soil particles. This refusal of the soil to absorb the water occurs either when the intensity of the rain is greater than the infiltrability of the soil surface, or when the rain falls on a surface that is partially or totally saturated by a water table.

These two types of runoff generally occur in very different environments. Once run-off has been triggered on a plot of land, erosion can take different forms, combining in time and space to give rise to diffuse and/or concentrated erosion.

It should be noted that runoff detachment (Fig. II.3) occurs when the frictional force of the water on the soil particles is greater than the shear strength of the soil, as shown in the figure below.

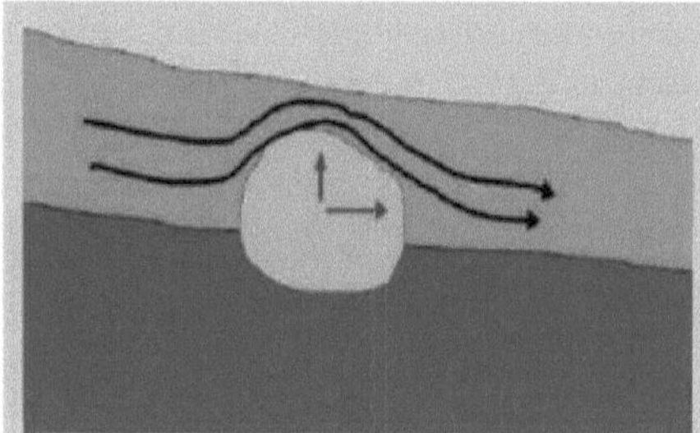

Figure II.3: Detachment by stream

b)   Transport

It is due to both raindrops (splash effect) and run-off water. Transport is therefore ensured by this water. However, it should be pointed out that transport by the splash effect is generally negligible except on steep slopes. On the other hand, run-off water is most responsible for transporting detached soil particles.

c)   Sedimentation (deposition)

The agent responsible for sëdimentation is run-off water. Particles drawn from the soil are deposited between the point of origin and the downstream point, depending on:

-   their size
-   their density
-   the transport capacity of the runoff or watercourse.

The particles are deposited in the following order:

-   Sand - Fine sand - Silt.

Clays and colloids are generally transported to the mouth of the river where they are deposited either after evaporation of the water or after flocculation.

II.4.3.2. Examples of damage caused by water erosion

1)   On site

-   Losses of soil and nutrients (Fig. II.4): the claws, fine gullies formed by water, particularly at the top of slopes, on the edge of tracks or in fields furrowed by ploughing, become gullies by widening due to the concentration of excessive runoff.
-   Losses of fertiliser and organic matter.
-   Destruction of soil structure.

-    Bank erosion (Fig. II.5) leads not only to the retreat of river banks, but also to an increase in the particulate load of river water [10].

Figure II.4: Stripping the soil surface

Figure II.5: Erosion of riverbanks by the flow of a river

2) Off-site

- Load rivers with suspended matter, causing an increase in water turbidity that alters the trophic balance.
- Muddy floods.
- Silting of river beds.

Figure II.6: Silting of river beds

11.5. Soil conservation issues in the Mediterranean basin [9,13].

Because soil degradation affects its ability to fulfil its ecological functions and the socio-economic uses that depend on it, soil conservation is a political and social issue. The measures that can be taken to meet this challenge are very diverse.

11.5.1. Agricultural issues

Agriculture is of major economic importance in most Mediterranean countries, and is accompanied by a large food processing industry, both upstream and downstream. It is therefore a major source of employment, especially in regions where it is still the most important activity. Preserving the Mediterranean's agricultural potential is therefore a key factor in maintaining the social and economic structure of our societies.

11.5.2. Territorial issues

The balance between the economic activities of a country's different regions depends, among other things, on the state of its agricultural and rural economy. Soil degradation can therefore become the cause of major territorial imbalances. The degradation of marginal land can lead to its abandonment, and to the migration of rural populations to the cities, which poses serious problems.

Chapter II:The phenomenon of erosion has economic and social implications in terms of regional planning and employment. In the context of sustained population growth in the southern and eastern Mediterranean countries, it is necessary to maintain rural areas capable of supporting a large population under satisfactory economic and social conditions.

11.5.3. Landscape issues

Agriculture plays a fundamental role in the management of Mediterranean natural resources, spaces and landscapes. However, the abandonment of certain areas that were once farmed (terraces, water management systems) can lead to irreversible degradation. In these cases, soil degradation is linked to the disappearance of certain typical Mediterranean ecosystems.

11.5.4. Environmental issues

Soils are a fundamental compartment of ecosystems: their degradation therefore generally has major impacts on all the other compartments, and seriously affects the composition and diversity of flora and fauna, as well as the water and nutrient cycles. Maintaining the biological diversity of the Mediterranean environment means incorporating ecological considerations into the planning of agricultural and urban development. In particular, we need to ensure that land artificialisation does not irreversibly destroy precious ecosystems.

11.6. Combating water erosion

The adoption of certain conservation practices, such as intercropping, the installation of water retention basins and the construction of terraces can help to reduce soil erosion. However, these measures can only be effective if the areas at risk of soil erosion are identified. A quantitative approach is therefore needed to better define these areas with a view to improving land management. The development and refinement of methods for analysing sediment sources and monitoring are important for determining areas where soil erosion and sediment production are critical [12,13].

To be effective, these control methods must be located in two distinct zones:

- A run-off zone,
- A sensitive area that accumulates precipitation.

Two aspects need to be taken into account:

- The agronomic (preventive) aspect, which covers cultivation techniques: soil cover, soil structure.
- The hydraulic aspect (remedial): various improvements (benches, terraces, etc.).

II.7. Partial conclusion

The phënomëne of erosion is part of the gëological revolution of the landscape under the effect of water and wind. The erosion of the earth's surface has continued throughout the ages. The movement, transport and departure of materials are natural phenomena that can be seen at any time and in any place. The most effective agents of erosion are rain, run-off and wind. The action of waves, frost and glaciers is limited to regions of restricted extent, but is significant in coastal areas and in glacial regions. The phenomenon of water erosion begins with the impact of the initial raindrop. It causes a great deal of environmental damage and will have major repercussions on our activities.

This research on erosion has enabled us to gain knowledge and a good understanding of the erosion phenomenon, in order to come up with the expected results and achieve the stated objective, which is to quantify water erosion in the Boussellam basin.

Estimation of erosion by RUSLE

Chapter III: Estimation of erosion using RUSLE

III.1 Introduction

Soil degradation is defined as a process that reduces the production potential of soils or the quality of natural resources. Water erosion is the main factor in soil resource degradation.

Dëtection of eroded areas, as well as evaluation of the factors controlling erosion and their characteristics, are complex tasks that can be solved by integrating several data sources (spatial data, field measurements and surveys, and satellite images) into geo-spatial processing systems, such as geographic information systems (GIS).

III.2 Erosion assessment methods

Several methods are used to assess soil erosion risks. Thus, two soil loss or erosion prediction modëles can be used given their universal adaptability. These are the R.U.S.L.E (Revised *Universal* Soil *Loss Equation)* modële and the Gravilovic (Erosion Potential Method) modële for predicting, respectively, soil loss at the slope level and sediment production at the subcatchment outlet level [14].

In this study, only the RUSLE model will be studied to quantify soil losses in the Boussellam catchment.

The approach used initially consists of detecting the factors triggering erosion and spatialising them using Landsat satellite images. Multi-temporal remote sensing data and GIS are used to assess and map each factor individually. Predictive modelling in a GIS environment offers an opportunity for erosion risk assessment. Data on erosion in relation to certain indicators are collected, calibrated and entered into a GIS database, after which they are spatially modelled to represent the risk of soil erosion in any selected landscape feature.

Individual layers have been created for each parameter of the RUSLE model and are then combined by a modelling procedure using ArcGIS software. All layers have ële projectedëes in UTM zone 31N using WGS 1984. The mëthodology adoptedëe
to carry out this study is shown in the figure below [7,15,19].

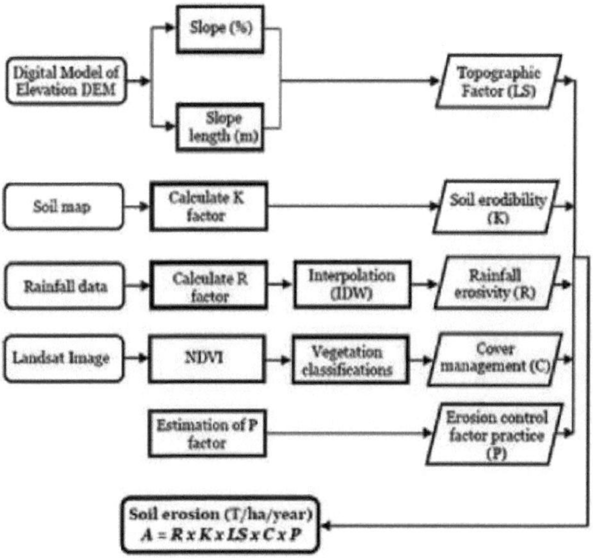

Figure III.1 : Methodology used to assess water erosion in the Boussellam catchment area

III.3. The RUSLE model

The modèle RUSLE [16] is the développement of liquation USLE [17]. The latter is widely adaptée to all ë scales. Basically, it has the advantage of providing long-term estimates of mean annual soil losses from small areas, and is considered a "good model" if the aim of modelling is to arrive at global estimates of soil erosion.

The RUSLE model takes the form of a model equation that uses erosion factors as inputs to estimate the average annual soil losses resulting from sheet and gully erosion. This model does not consider erosion processes such as détachment, transport and dëpôt [18]. It retains the same form as the equation usedëe in the USLE model, and that is presented as the following relationship.

$$A = R * K * LS * C * P \qquad \ldots\ldots\ldots \text{(III.1)}$$

Or :

A: is the annual soil loss rate (t/(ha.year))

Chapter III:Estimation of erosion by RUSLE R: is the rainfall erosivity factor, it corresponds to the annual average of the sums of the products of the kinetic energy of the rain by its intensity in 30 consëcutive minutes ((MJ -mm)/(ha -h- yr))

K: is the erodibility of soils, and depends on the granulometry, quantity of organic matter, permeability and structure of the soil ((t.h.ha)/(MJ.ha.mm))

LS: is a dimensionless factor representing the slope (S in %) and the length of the slope (L in m)

C: is a dimensionless factor representing the effect of plant cover

P: is a dimensionless factor, a ratio that takes account of anti-erosive cultivation techniques, such as contour ploughing.

The aim of this development is :

-   Extending the range of variation of the R factor to new areas
-   Development of a periodically variable soil erosion susceptibility term (K) and alternative methods for estimating (K)
-   A new method for calculating the (C) factor
-   Other formulas for estimating the (LS) factor, taking into account variable topography.

III.4. Results and discussion

III.4.1. _Calculation of the rainfall aggressivity factor (R)_

Estimation of the factor (R) according to Wischmeier's formula requires knowledge of the kinetic energies (E_c) and the average intensity over 30 minutes (I_{30}) of the raindrops in each shower. These are given by the following empirical formula [18].

$$R = K. \, E_c \,.I_{30}$$

(K) being a coefficient dependent on the measurement unit system.

The only precipitation data available from stations in or near the basin are monthly and annual averages.

Authors such as _Kalman (1967), Arnoldus (1987) and Rango & Arnoldus (1987)_ [7,9,15,19] have developed alternative formulae that use only monthly and annual precipitation to determine the factor (R).

$$\log R = 1.74.\log\Sigma\ (P_i^2/P) + 1.29 \quad \ldots\ldots\ldots \quad (\text{III.2})$$

Or :

P_i: Average monthly precipitation (mm)

P: average annual precipitation (mm).

The results of calculating the erosiovite coefficient (R) are shown in Table (III.1).

Table III.1: Average monthly and annual precipitation values in (mm) and R for the period (1970 -2017) [4,6]

N°	Code	Seven	Oct	Nov	Dec	Jan	Feb	Mar	Apr	May	June	July	August	P$_{an}$	R
1	50403	24.81	34.25	38.5	47.42	40.62	41.56	38.69	42.86	39.89	17.46	6.38	8.97	390.75	55.015
2	50501	24.69	31.53	36.47	75.93	68.58	49.17	52.17	40.05	27.45	14.34	4.36	6.99	400.58	72.83
3	50503	26.61	34.3	42.86	70.49	66.83	55.2	52.41	46.29	35.89	16.71	5.59	8.68	452.74	70.14
4	50603	27.78	32.42	34.35	26.47	31.27	30.07	31.78	34.73	38.66	20.56	7.9	10.13	346.4	45.70
5	50606	32.95	36.72	37.18	63.73	48.28	39.9	43.92	40.81	43.43	21.81	8.35	11.11	420.32	61.99
6	50607	33.29	37.74	39.38	54.83	49.55	41.49	45.86	43.36	44.3	22.06	8.39	11.22	432.45	60.53
7	50608	32.81	36.88	39.5	60.29	52.54	46.1	51.62	46.77	43.59	20.85	7.87	10.75	441.99	64.47
8	50614	28.3	34.36	37.18	40.26	38.19	33.52	33.89	37.62	40.01	21.32	8.02	10.43	370.9	51.39
9	50703	27.52	47.98	57.82	98.92	92.93	84.26	65.88	57.64	43.85	15.81	4.81	7.84	612.82	88.38
10	50706	36.07	46.51	52.04	81.51	70.6	61.89	62.65	52.45	49.13	20.93	7.89	11.22	555.94	76.48
11	150707	29.51	31.73	39.07	50.84	39.87	36.72	42.53	37.81	39.8	20.59	8.32	9.06	385.85	55.76
12	50802	31.79	39.72	48.14	80.6	61.97	57.97	59.23	55.47	45.3	19.79	7.16	10.37	500.62	75.58
13	150904	26.31	32.83	43.04	62.16	56.59	42.31	42.82	38.37	29.24	8.14	4.49	8.55	394.85	61.81
14	51002	25.07	44.74	52.38	79.57	80.95	70.22	55.25	48.55	36.31	14.16	3.91	6.37	532.52	76.76
15	51003	37.57	63.96	82.2	126.63	117.8	104.15	81.45	59.61	36.99	15.84	3.26	5.42	735.29	107.26
16	51007	32.85	69.44	87.4	149.13	142.5	116.78	98.94	75.32	47.48	16.31	4.19	7.56	834.52	122.51
17	70103	43.06	26.73	34.33	45.62	52.57	43.02	43.17	45.29	37.05	25.12	4.33	15.91	416.2	58.96
18	100106	27.43	54.41	60.91	115.44	98.83	112.01	94.74	94.73	49.35	15.19	6.64	5.4	735.08	105.89
19	1510	48	89	98	122	113	111	89	74	42	14	7	11	818	110.80
20	70102	30	32	39.5	50.8	40.1	37	42.6	38	40	20.4	8.89	9.8	389.09	55.90
21	150602	29.51	31.73	39	50.84	39.87	36.72	42.53	37.81	39.8	20.59	8.32	8.5	385.22	55.79
22	51111	27.93	34.66	36.88	40.12	37.11	33.52	33.89	37.62	39	21.5	8	9.43	359.66	51.93
23	51011	23.7	19	17.6	24	20.56	18.75	21.55	25.75	30.8	9.1	8.25	10	229.06	36.82
24	50901	26.6	28.8	35.27	42.85	47.47	32.47	28.67	33.3	29.8	7	5.4	4.19	321.82	51.81
25	150801	33.12	26.89	31.3	38.47	36.56	27.24	31.12	39.59	38.54	15.94	3.97	11.44	333.88	50.18

The erosivity map synthëtisëed from the spatialisation of the hydrological stations shows that the value of the R factor varies from 36.82 to 122.51 MJ.mm/ha.h.an (Fig. III.2). The highest ëlevë values are recorded in the north of the basin, while the lowest values are recorded in the south. In fact, R values undergo an increasing gradient from the south to the north of the catchment (from upstream to downstream).

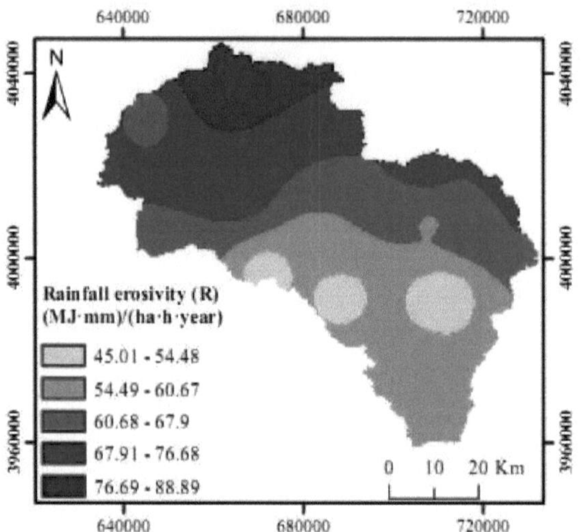

Figure III.2: Map of the (R) factor

1.1.1. % of the total surface area of the catchment is subject to high climatic aggressiveness corresponding to an R class greater than 67.91 MJ.mm/ha.h.yr. R values vary between 44 and 140 MJ.mm/ha.h.yr, with an average of 65.33 MJ.mm/ha.h.yr. The lowest R values in the class (45.01 to 54.48 MJ.mm/ha.h.yr) are in the south of the catchment, while the highest values (> 76.68 MJ.mm/ha.h.yr) are in the north of the catchment and in mountainous areas. These values show that the Boussellam catchment is subject to aggressive rainfall.

1.1.2. Determination of the topographical factor (LS)

The topographical factor combines the effects of slope length (L) and slope steepness (S) on erosion [16]. The length of the slope determines the speed of runoff and particle transport increases with the length of the plot.

The equations used to calculate the (LS) factor are :

- **F = (Sin("slope_degree" * 0.01745) / 0.0896) / (3 * Power(Sin("slope_degree" * 0.01745),0.8) + 0.56))** .... (III.3)

- **m = "F" / (1 + "F")** .... (III.4)

- **L = Power((("FlowAcc" + 625), ("m" + 1)) – Power("FlowAcc", ("m" + 1)) / Power(25,("m" +2)) * Power(22.13,"m"))** .... (III.5)

  - **S = Con(Tan("slope_degree" * 0.01745) < 0.09,(10.8 * Sin("slope_degree" * 0.01745) + 0.03),(16.8 * Sin("slope_degree" * 0.01745) - 0.5))** .... (III.6)

__Chapter III:Estimation of erosion by RUSLE__ The values obtained for the LS factor have ëtës grouped into five classes of values (Fig. III.3). The distribution of the topographic factor (LS) shows that 99.99 % of the catchment area falls within the class 0.03 to 5. Consequently, most of the catchment is subject to a high risk of erosion.

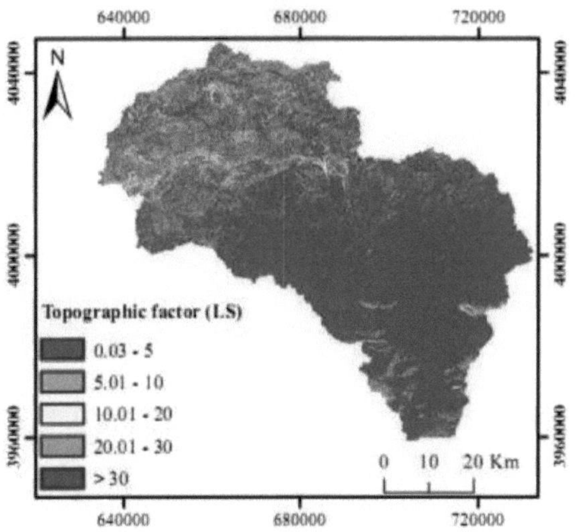

Figure III.3: Factor map (LS)

1.1.3. Determination of soil erodibility factor (K)

The hydrodibility of a soil is expressed by its inherent resistance to the detachment and transport of particles by water. The ërodΓbilitë factor (K) of a soil expresses its sensitivity to water erosion and depends on its intrinsic properties, i.e. its texture, structure and permeability. It is determined by the following relationship [16,17].

$$100K = 2.1 * M^{1.14} * 10^{-4} (12 - a) + 3.25 * (b - 2) + 2.5 * (c - 3) \quad .... \quad (III.7)$$

Or :

M = (% Silt) x (100 - % Clay)

a: percentage of organic matter **b**: structure code **c**: permeability code.

After downloading the organic matter, sand and clay maps from the website (https://soilgrids.org/#!/? layer=ORCDRC M sl2 250m&vector=1), we calculate the parameter

(M) by the formula (Ш.8), and then we derive codes (b) and (c) using the
diagrams below [8].

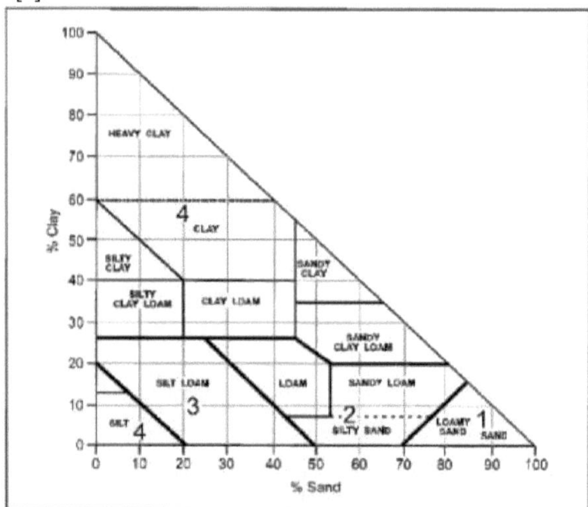

Figure III.4 : Structure code (b)

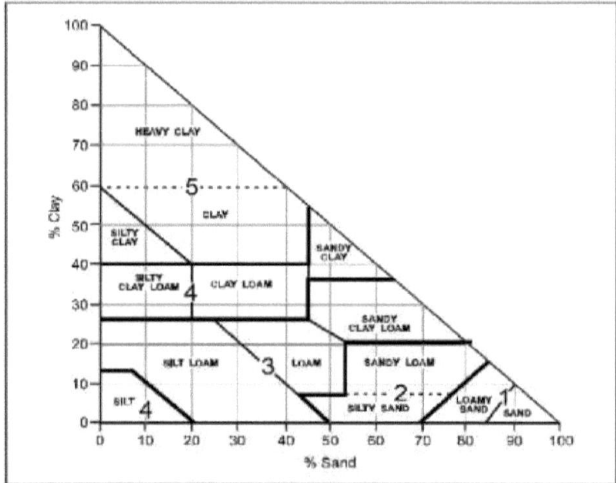

Figure III.5: Permeability code (c)

Figure (Ш.6) shows a map of the ërodibilitë factor (K) for the Boussellam catchment.

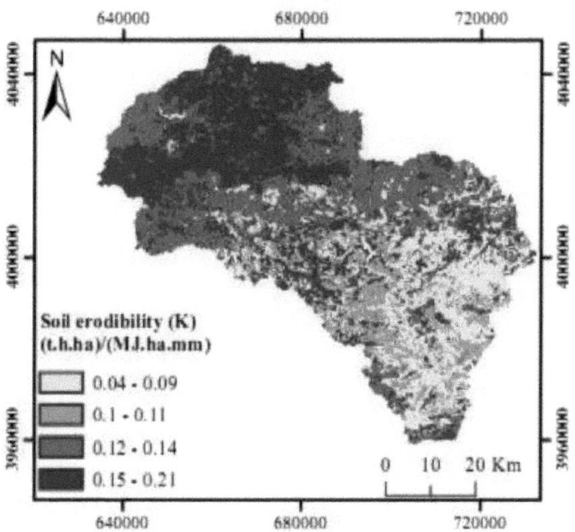

Figure III.6: Map of the (K) factor

The results show that most of the soils in the catchment area (67%) have a low ërodibilitë (< 0.2). These regions are characterised by an impermeable lithology comprising karstic and sandy facies. Only 33% of the surface area of the catchment has a ërodibilitë ëlevëe greater than 0.2, located mainly in the middle and south of the catchment. The average value of the factor (K) is 0.11 (t.h.ha)/(MJ.ha.mm).

111.4.4.   Anti-erosive practice (P)

The (P) factor (varying from 0 to 1) designates antiërosive cultivation practices. These practices proportionally affect erosion by modifying the flow pattern or direction of surface runoff, and by reducing the quantity and speed of runoff.

The cartographic representation of the slope map of the Boussellam basin (*Fig. 1.13- Chap. I*) shows that slopes below 19° make up 84% of the total surface area of the catchment, and slopes above 69° make up almost 4% of the total surface area. The very high percentage of gentle slopes proves the rarity of anti-erosion practices in the catchment [17].

The values obtained for this factor are 26.6% for P = 1, 47.47% for P = 0.9 and 25.92% for P = 0.55. The average value of the factor (P) is 0.77; it is close to 1, which confirms that there are no erosion control practices in the basin.

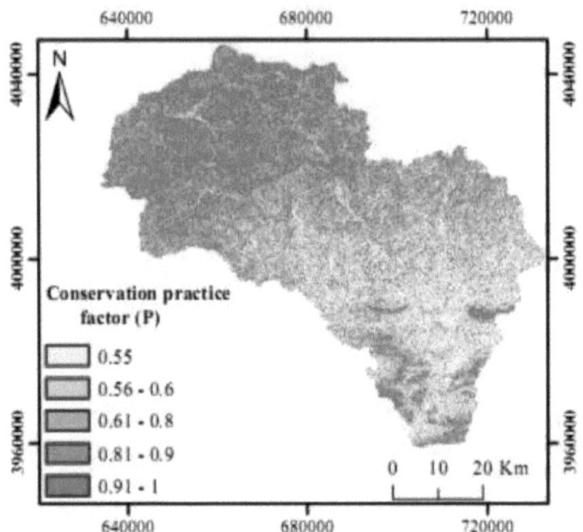

Figure III.7: Map of factor (P)

111.4.5.   Land use (C)

The factor "vëgëtal cover" (C) is the second most important factor controlling the risk of soil erosion. This dimensionless factor can be estimëed from the normalisedë difference vëgëtation index (NDVI), dëterminë of satellite imagery. In fact, the value of this factor depends on the nature of the vegetation and the percentage of vegetation cover [15,20].

The satellite images were collected from the USGS (http://landsat.usgs.gov/), on the Landsat 5 and 8 satellites, with a resolution of 30*30 pixels.

The two images were taken on clear days (not cloudy or snowy), and consist of two scenes:

For the dry period, the acquisition date is 06/07/2015 and 13/07/2015.

For the wet period, the acquisition date is 18/03/2016 and 25/03/2016.

a)   Principle of remote sensing

The principle of remote sensing is similar to that of human vision. Remote sensing is the result of the interaction of three fundamental elements:

- The target is the element or portion of the Earth's surface observed by the stellite.
- The energy source that illuminates the target by emitting an electromagnetic wave.
- The sensor or remote sensing platform measures the energy reflected by the target.

b)   What is NDVI?

NDVI is constructed from the red (R) and near infrared (NIR) channels. The normalised vëgëtation index highlights the difference between the visible red band and the near infrared band [10].

$$NDVI = (NIR - R) / NIR + R) \qquad \cdots\cdots (III.8)$$

$$NDVI = (band4 - band3) / (band4 + band3) \text{ pour Landsat TM5 et TM7}$$

$$NDVI = (band5 - band4) / (band5 + band4) \quad \text{pour Landsat LC08}$$

This index is sensitive to the vigour and quantity of vegetation. NDVI values range from -1 to +1, with negative values corresponding to surfaces other than vegetation cover, such as snow, water or clouds, for which the red reflectance is greater than that of the near infrared. For bare soil, the red and near-infrared reflectances are of roughly the same order of magnitude, so the NDVI has values close to 0.

Vegetation formations have positive NDVI values, generally between 0.1 and 0.7. The highest values correspond to the densest cover.

c) Steps in determining the NDVI index

NDVI is a number that indicates the probability that the area observed contains vegetation. It is generally used in remote sensing. The NDVI can be calculated from the amount of light that the observed area reflects in the near infrared and red regions of the light spectrum. These instructions will provide the steps for calculating NDVI.

-       Examine the importance of spectral reflectance in the near infrared region. Plants tend to reflect light in this region because there is not enough energy for photosynthesis. A high reflectance factor in this region therefore increases the probability that the area observed will contain vegetation.

-       Examine the importance of spectral reflectance in the red region. Plants tend to absorb light in this region, as there is enough energy for photosynthesis, but not so much. A low reflectance factor in this region therefore increases the probability that the area observed will contain vegetation.

-       Define NDVI mathematically as NDVI = (NIR - RED) / (NIR + RED). Higher NDVI values will result in a high NIR value and a low RED value, indicating vegetation.

-       Examine the NDVI range.

      -Interpreting NDVI values. The dense vëgëtation must have an NDVI value between 0.3 and

0,8. Clouds and snow will have negative NDVI values.

Vegetation classification is performed using the following thresholding conditions [15, 21]:

NDVI < -0.1 : Water

-0.1 < NDVI < 0.15 : Bare soil

0.15 < NDVI< 0.25 : Sparse vegetation

0.25 < NDVI< 0.4 : Moderately dense vegetation

NDVI > 0.4 : Dense vegetation

In order to estimate the values of the C factor in the study area, polynomial regression between C and NDVI was used. These values are taken from the experimental diagram shown in figure (III.8) [22].

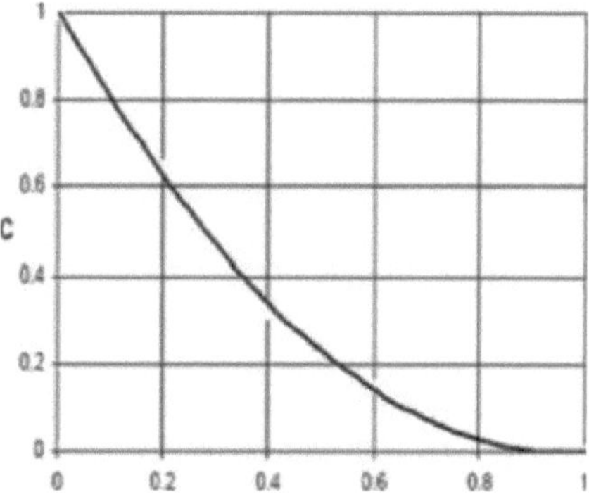

Figure III.8: Experimental diagram for estimating factor (C) [22].

The equation of the polynomial curve resulting from the regression is: $C = 1.1119 * (NDVI)^2 - 2.0976$

$$C = 1.1119 * (NDVI)^2 - 2.0976 * (NDVI) + 0.9944 \qquad \ldots\ldots\ldots (III.9)$$

Figure (III.9) shows the raster mosaic covering the Boussellam basin (July 2015).

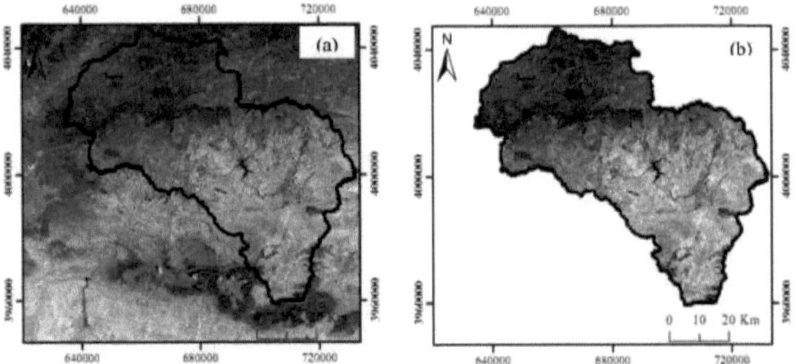

Figure III.9 : Preparation of satellite images. a) *Mosaic of two scenes covering the B.V (LC08_Band 4_July 2015).* **b)** *Extract of images covering the B.V of Boussellam*

Figures (III.10) and (Ш.11) show NDVI values in the Boussellam basin.

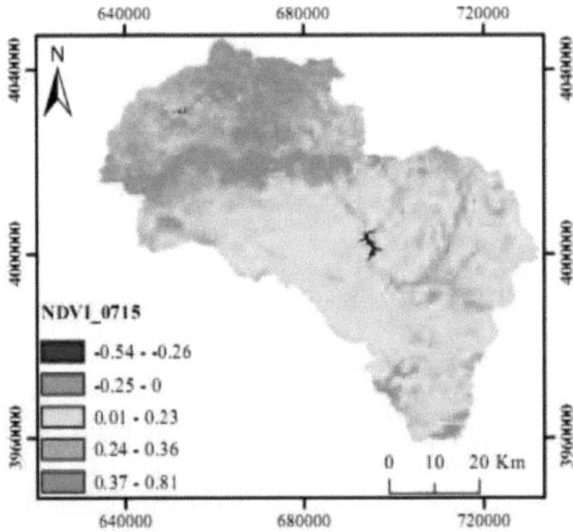

Figure III.10: Map of NDVI values (July 2015)

The NDVI study showed that there is a variation in the vegetation cover between the wet and dry periods. The most ëlevëes values (0.3 to 0.86) corresponding to a dense vëgëtation are located in the North for both përiodes. Vëgëtation gradually decreases towards the south. This area is considered to be a dëgradëe zone.

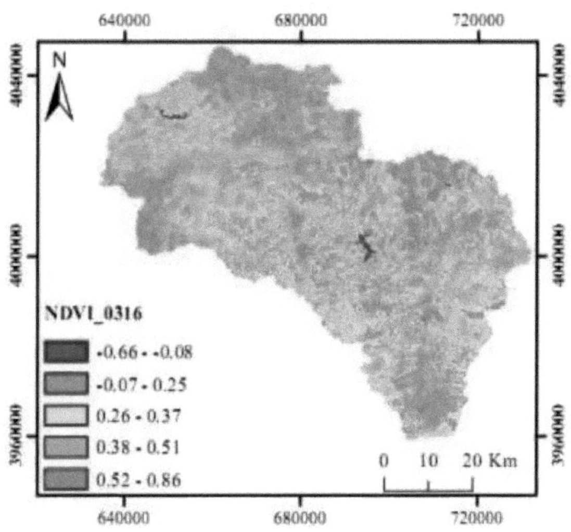

Figure III.11: Map of NDVI values (March 2016)

The maps of factor C (Figs. Ш.12 and III.13) are gënërëes using polynomial liquidation (Eq. Ш.9). For negative NDVI values, the C factor is
considersërë as nil because there is no vegetation cover (water) on these surfaces.

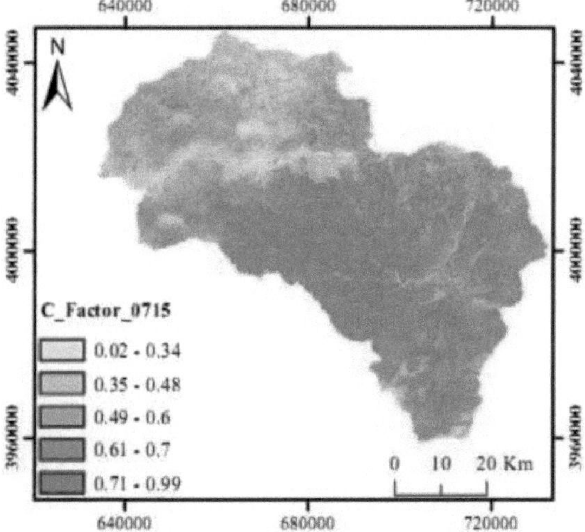

Figure III.12: Factor C map (July 2015)

41

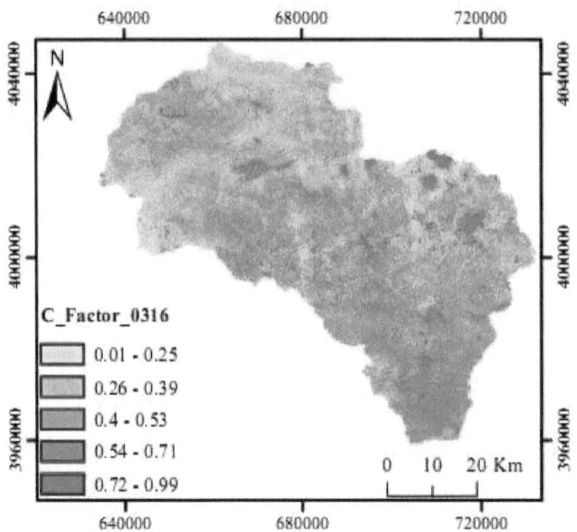

Figure III.13: Factor C map (March 2016)

The results obtained show that more than 80% of the area of the basin has very low vëgëtal cover for both periods, and only 7% (for July 2015) and 19% (for March 2016) of this area is well protëgëes with C < 0.3.

Most of the basin is characterised by values of the factor (C) between 0.5 and 0.7. These areas are vëgëtations modërëes. The second largest class is made up of values ranging from 0.4 to 0.6. The regions in which this second class is located are mainly in the north and centre of the catchment area, with varied vegetation. This can be explained by the dominance of degraded rangelands and arable areas which are considered to be very sensitive to erosion.

III.4.6. Annual soil loss rate (A)

By superimposing the maps of the factors involved in soil water erosion, it was possible to obtain a map of soil losses at all points in the Boussellam catchment area (Figs. III.14 and III.15). Soil losses range from 0.0058 to 578.39 t/ha/yr for July 2015, with an average of 258.06 t/ha/yr, and from 0.0013 to 1164.89 t/ha/yr for March 2016, with an average of 190.92 t/ha/year, reflecting the high erosion of the basin. 66% of the basin area for March 2016 and 56% of the basin area for July 2015 had losses below the tolerance threshold (< 7 t/ha/year).

Low levels of erosive risk are found in the southern part of the basin. The land in these areas is virtually flat (slope less than 11.5%).

The highest values are found in the northern part of the catchment area. The ërosive risk is important in these regions because of the steepness of the slopes, but also because of the high temperatures.

R factor values despite the presence of dense vëgëtation in these areas.

42

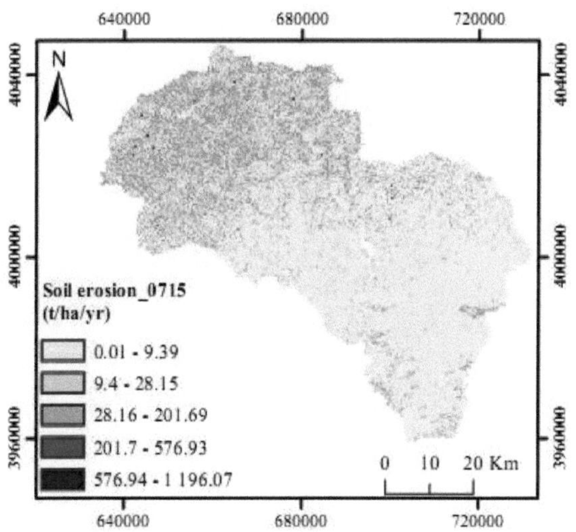

Figure III.14: Annual soil loss rate (July 2015) in (t/ha/yr)

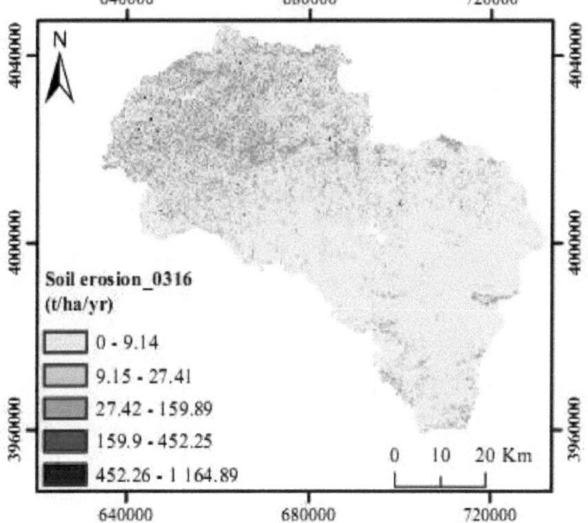

Figure III.15: Annual soil loss rate (March 2016) in (t/ha/year)

111.5. Partial conclusion

The results obtained show that the soils in the Boussellam catchment area are affected by several factors that favour erosion, namely the steepness of the slopes, the low vegetation cover and the erodibility of the soils. They also indicate that the Boussellam catchment is subject to heavy erosion, with losses ranging from 27 to 1196 t/ha/year, affecting 34% of the area in March 2016 and 43.34% in July 2015.

General conclusion

General conclusion

The literature review provided us with a good understanding of the topography, climatic conditions, morphomëtric characteristics and hydrographic network of the Boussellam catchment.

In addition, the use of GIS and tëlëdëtection made it possible to apply theonic mëthods to solve spatial analysis problems and to process gëographic data.

The results obtained show that the soils in the Boussellam catchment are affected by several factors that favour erosion, namely the steepness of the slopes, the low vegetation cover and the erodibility of the soils. They also indicate that the catchment is subject to severe erosion, with losses ranging from 27 to 1196 t/ha/year, affecting 34% of the area in March 2016 and 43.34% in July 2015. This serious situation in the north of the catchment is exacerbated by a number of factors that combine to accelerate erosion: steep slopes, worryingly aggressive rainfall, alarming degradation of plant cover and highly erodible soils.

The study showed that the application of the RUSLE model can provide important assistance to decision-makers in planning interventions to combat the damage caused by erosion.

A number of issues need to be addressed:

- Seasonal evolution of NDVI values over a long period using remote sensing.
- Soil analysis to validate K factor results.
- Advanced studies in geology, geophysics and hydrogeology.

References

References

[1] A. Amour, Caracterisations des inondes pluviales des sous bassins de la Soummam. *Memoire de Magister en Hydraulique generale*, Universite de Bejaia, (2010).

[2] S. Djenba, Influence des parametres geologique, geomorphologique et hydrogeologique sur le comportement mecanique des sols de la wilaya de Setif. (Algerie). *These de Doctorat en Sciences*, Universite de Biskra, (2015).

[3] D. Sersoub, Amenagement et sauvegarde de la biodiversite de la vallee d'Oued Boussellam (Setif). *Memoire de Magister*, University of Setif, (2012).

[4] M. Smaili and A. Touati, Contribution a la caracterisation des eaux de cinq sources dans le bassin versant de Boussellam, Sud-est de Bejaia-Algerie. *Memoire de Master en Toxicologie Industrielle et Environnementale*, Universite de Bejaia, (2018).

[5] M. Durand Delga, Mise au point sur la structure du Nord-Est de la Berberie. *Publ. Serv. Carte geol. Algerie, N. S., Bull.* 39, (1969) 89-131.

[6] N. Ben Hamiche, Contribution a l'etude de l'influence climatique, lithologique et anthropique sur la variation des parametres physico-chimiques des eaux d'un aquifere du Nord-est algerien : Cas de la basse Soummam, Bejaia. *These de Doctorat en Sciences*, Universite de Bejaia, (2015).

[7] Ch. Allili, B. Laignel, N. Adjeroud, H. Bir, K. Madani, Particulate flow at the mouth of the Soummam watershed (Algeria). *Environmental Progress & Sustainable Energy*, 35 (1) (2015) 204-211.

[8] M. Baaidja, A. Bachiri, Evaluation des pertes en sols et leur identification dans le bassin versant d'Oued Azzerou (BBA). *Memoire de fm d'etudes, Master en Hydraulique urbaine*, Universite de M'sila, (2018).

[9] L. Bouhali, Cartographie des risques d'erosion des sols dans le bassin versant de la Soummam. *Memoire de fin d'etudes, Master en Ouvrages hydrauliques et Amenagement*, Universite de M'sila, (2016).

[10] M.M. Benachour, Etude de l'erosivite des pluies sur le bassin de la Soummam par le biais de SIG et teledetection. *Memoire de fin d'etudes, Master en Hydraulique urbaine*, ENSH de Blida, (2016).

[11] B. Heusch, L'erosion du Pre Rif occidental: une etude quantitative de l'erosion hydrique (1988).

[12] E. Roose, Conventional erosion control based on water erosion processes and factors (1994).

[13] R. Neboit, L'Homme et l'erosion. Published by Presses Universitaires Blaise Pascal, Nature & Societes collection, (2010).

[14] M. Yjjou, Modelisation de l'erosion hydrique via le SIG et l'equation universelle de perte en sol au niveau du bassin versant d'Oum Er Rbia. *Memoire de fin d'etudes, Master en Sciences des Sols et Environnement*, Universite Moulay Ismail, Meknes (Maroc), (2009).

[15] Y. Sahli, E. Mokhtari, B. Merzouk, B. Laignel, C. Vial, K. Madani, Mapping surface water erosion potential in the Soummam watershed in Northeast Algeria with RUSLE model. *Journal of Mountain Science*, 16 (7) (2019) 1606-1615.

[16] K.G. Renard, G.R. Foster, G.A. Weesies, D.K. McCool, D.C. Yoder, Predicting rainfall erosion by water: A guide to conservation planning with the revised universal soil loss equation (RUSLE). *Agriculture handbook*, USDA, 703 (1997).

[17] W.H. Wischmeier, D.D. Smith, Predicting rainfall erosion losses - a guide to conservation planning. *Agriculture handbook*, USDA, 537 (1978).

[18] D.K. McCool, L.C. Brown, G.R. Foster, Revised slope steepness factor for the Universal Soil Loss Equation. *Transactions of the American Society of Agricultural Engineers*, 30 (1987) 1387-1396.

[19] A. Sadiki, B. Saidati Bouhlassa, A. Jamal, F. Ali, J.-J. Macaire, Utilisation d'un SIG pour revaluation et la cartographie des risques d'erosion par l'Equation universelle des pertes en sol dans le Rif oriental (Maroc) : cas du bassin versant de l'oued Boussouab. *Bulletin de l'Institut Scientifique, Rabat, Section Sciences de la Terre*, 26 (2004) 69-79.

[20] R. Kalman, Le facteur climatique de l'erosion dans le bassin du Sebou. *Report by the Ministry of*

Agriculture, Morocco, (1967).

[21] S. Toumi, Application des techniques nucleaires et de la teledetection a l'etude de l'erosion hydrique dans le bassin versant de l'Oued Mina. *These de Doctorat en Sciences,* ENSH de Blida, (2013).

[22] I.Z. Gitas, K. Douros, C. Minakou, G.N. Silleos, Multi-temporal soil erosion risk assessment in N. Chalkidiki using a modified USLE raster model. *EARSeL eProceedings*, 8 (1) (2009) 40-53.

Printed by Books on Demand GmbH, Norderstedt / Germany